Ahmed ISSOUFOU IMADAN

Manual for dimensioning a PV solar pumping system

Ahmed ISSOUFOU IMADAN

Manual for dimensioning a PV solar pumping system

Demonstration of a sizing method for a solar PV pumping system: Application to a case study

ScienciaScripts

Imprint
Any brand names and product names mentioned in this book are subject to trademark, brand or patent protection and are trademarks or registered trademarks of their respective holders. The use of brand names, product names, common names, trade names, product descriptions etc. even without a particular marking in this work is in no way to be construed to mean that such names may be regarded as unrestricted in respect of trademark and brand protection legislation and could thus be used by anyone.

Cover image: www.ingimage.com

This book is a translation from the original published under ISBN 978-620-6-71363-0.

Publisher:
Sciencia Scripts
is a trademark of
Dodo Books Indian Ocean Ltd. and OmniScriptum S.R.L publishing group

120 High Road, East Finchley, London, N2 9ED, United Kingdom
Str. Armeneasca 28/1, office 1, Chisinau MD-2012, Republic of Moldova, Europe
Printed at: see last page
ISBN: 978-620-7-66689-8

General introduction

Conventional energy sources such as nuclear power and fossil fuels (coal, oil and gas) are derived from limited stocks of materials extracted from the earth's subsoil. Each of these sources, when used, has more or less significant long-term consequences for the environment, which are tending to be better controlled: atmospheric pollution, climate change, radioactive contamination.... .... Renewable energy sources, on the other hand, are clean and unlimited, their use does not pollute the atmosphere and they do not produce greenhouse gases such as carbon dioxide and nitrogen oxides, which are responsible for global warming. The sources of these energies can be: wind (wind turbines), water (hydroelectricity), vegetation (biomass), the sun, which can be used in two ways:

- its heat can be concentrated to heat domestic hot water, buildings, dryers, or a circulating fluid to produce electricity via an alternator or dynamo-electric: this is solar thermal energy.
- its light can be converted directly into electrical current using the photovoltaic effect (which will be studied in this thesis).

Since the discovery of photovoltaic energy, the use of solar energy has become one of the most promising applications: in the developed countries where photovoltaic solar energy is most commonly used at present, it is usually used to supplement electrical energy for domestic use in urban areas, and solar panels are placed on the roofs of homes.

In developing countries, most of which are equatorial or tropical, and therefore very sunny, and which have many localities where the electricity grid is absent, photovoltaic energy should therefore have no competition. This is even more true in the case of water pumping applications in Sahelian countries such as Niger, where maximum demand for water is matched by optimum availability of solar resources.

However, it should be noted that water pumping in the Sahelian states is generally carried out with the aim of satisfying the drinking water needs of the population, farmers and livestock in regions where no water distribution network is planned or available.

Theoretical and experimental studies [8] have shown that solar water pumping for drip irrigation is the best compromise for developing agriculture in Sahelian countries, because of the water savings that this irrigation system brings.

However, since a water pumping system is made up of a large number of operating components, each of the constituent elements must be correctly sized. To this end, the study presented in this dissertation focuses on the sizing of certain components and the selection criteria for the other components of a solar photovoltaic pumping system in order to design an efficient, reliable and economically profitable system.

Work objectives

General objective

Designing a high-performance, cost-effective photovoltaic solar pumping system

Specific objectives

- Determine the energy produced by the system as a function of the amount of sunshine on site
- Determine the actual flow rate of the installed pump
- Sizing the photovoltaic array
- Carry out an economic study of the system installed

This study will be structured in four chapters (4) divided as follows:

-The first chapter presents the host organisation and the aim of the work.

-The second chapter will first be devoted to general information and a presentation of the components of a photovoltaic solar pumping system.

-The third chapter will then propose a model for selecting and sizing a solar pumping system, which will be applied to a real case.

-The final chapter will focus on verifying the performance of the installed system, in order to assess the economic profitability of the system in contrast to the use of conventional energy sources, and will conclude with recommendations.

This study will end with a general conclusion.

CHAPTER 1: PRESENTATION OF THE RECEPTION STRUCTURE

1.1. Presentation of the French National Solar Energy Centre (CNES)

1.1.1. History

The Centre was created in May 1965, originally under the name "Office de l'Energie Solaire", under the acronym "ONERSOL" and with the status of a Public Industrial and Commercial Establishment (EPIC). This status gave the organisation the task of carrying out both :

- ✓ Research and development of systems based on solar energy;
- ✓ And testing, manufacturing and marketing of prototypes selected for their functionality.

With the creation of a solar equipment production plant, particularly for solar water heaters, the entity enjoyed resounding success both nationally and sub-regionally. Since the factory was set up in 1976 and 1989, ONERSOL has produced and marketed :

- ✓ 570 water heaters of 200 litres/day ;
- ✓ 100 filters (aluminium tanks) for the Ministry of Public Health (MSP);
- ✓ 100 aluminium ice boxes for the M.S.P;
- ✓ 1,572 square metres of flat-plate thermal collectors.

In the field of Research and Development, the Office can pride itself on having designed and developed technologies that were already anticipating the future of solar energy in Africa, and indeed the world. These include technologies such as flat-plate collectors with multiple glazing, and solar concentrators for use as heat sources in thermal engines.

In 1982, the Office had to be restructured, in particular due to the bankruptcy of the Manufacturing and Marketing Section. It was therefore decided to transform the Research Section into an EPA under the name of "Centre National de Recherche en Energies Nouvelles et Renouvelables (CNRENR)" and the Manufacturing and Marketing Section into a semi-public company, the "Société Nigérienne d'Energies Nouvelles (SONIEN)".
Unfortunately, the texts prepared for this purpose were not adopted by the competent authorities and the restructuring in question was not effective, even with the disappearance of the Manufacturing Section. In fact, the latter was liquidated in 1985 with the dismissal of its staff.
It was in this situation of unfinished reform and an unfavourable economic climate that

ONERSOL continued to operate as best it could until 1997, when it was restructured once again with the adoption of law no. 97-024 of 08 July, which created, in place of ONERSOL, the Centre National d'Energie Solaire (CNES), under the status of a Public Administrative Establishment (EPA) placed under the authority of the Ministry in charge of Energy, whereas ONERSOL was under the authority of the Ministry in charge of Higher Education and Research.

1.1.2. CNES missions

Following the change of status, the new structure has been given the following responsibilities:

- ✓ To carry out research into the use of renewable energies, particularly solar energy, and to disseminate the results;
- ✓ Participating in prospective and diagnostic studies on the use of renewable energies in all sectors of the national economy;
- ✓ Participating in training in the field of Renewable Energies.

1.1.3. The CNES organisation

The Centre is administered by a Board of Directors whose members are appointed for a renewable term of three (3) years by order of the Minister responsible for Energy.
In addition to the Executive Board, which is the governing body, the Centre is structured into directorates, including :

- ✓ The Research & Development Department ;
- ✓ Engineering Department ;
- ✓ Studies, Monitoring & Evaluation Department;
- ✓ Administration & Finance Department.

1.1.3.1. Engineering Department

I did my work placement in this department. Its tasks include :

- Engineering renewable energy projects (preparing programmes, identifying projects, studying them)
- Quality and conformity control of renewable energy equipment introduced in Niger

- The launch and supervision of work to install equipment powered by renewable energy.
- Ex-post monitoring and evaluation of renewable energy projects.
- Market studies; prospective studies (prospective studies on the development of renewable energies in a sustainable energy context).
- Technical assistance to government bodies and the private sector implementing renewable energy programmes and for equipment maintenance.

1.1.4. Infrastructure

The Centre has :

- ✓ A laboratory complex comprising 15 offices and laboratories,
- ✓ A welcoming city ;
- ✓ Two (2) experimental and measurement sites;
- ✓ A workshop for manufacturing equipment prototypes.

CHAPTER 2: GENERAL INFORMATION ON PHOTOVOLTAIC SOLAR PUMPING

Introduction

Many regions in developing countries suffer from a lack of water, a vital element for people, livestock and agriculture. The water that is traditionally drawn (using skins or pumps powered by traditional energy) is insufficient and often of poor quality. There are also maintenance and fuel supply problems when diesel generators are used.

As developing countries generally have a lot of sunshine, photovoltaic pumping, which can deliver clean, abundant water without intervention, quickly emerged as a solution to the problem of water supply in rural areas. Two solutions are often used:

- Battery-free or "solar-powered" pumping (the case studied here);
- Battery-powered pumping

The most widespread system is the solar pumping system.

In this chapter we will study all the essential elements that make up a photovoltaic solar pumping system, namely the photovoltaic generator, the motor pump unit, the conversion elements and the storage section.

2.1. Diagrams and description of the principle

Photovoltaic solar panels produce electrical energy in the form of direct current, which is converted by a static converter to power a submersible or floating motor-driven pump unit. The pump unit consists of a single-, two- or three-phase AC motor or an electronically commutated DC motor coupled to a multi-stage centrifugal pump or a positive-displacement pump, depending on the flow rate required.
The centrifugal pump transmits the kinetic energy of the motor to the fluid by means of a rotating movement of impellers, while the positive displacement pump transmits the energy of the motor by means of a helical movement that literally propels the water to the surface. The proposed systems consist of photovoltaic modules mounted on a support frame that is tilted according to the latitude of the site in order to optimise photovoltaic production, or rotates according to the path of the sun. The system is completed by a surface-mounted static converter that converts the direct current produced by the solar field into alternating or direct current to power the motor coupled to the pump. A reservoir can be placed at the

pump outlet to store water for use even when the sun is not shining. Pumps can also be operated by adding batteries. In this case, a battery bank is added to the installation, along with a charge controller that manages battery charging. The solar panels charge the batteries during the day and the pump can be used at night.

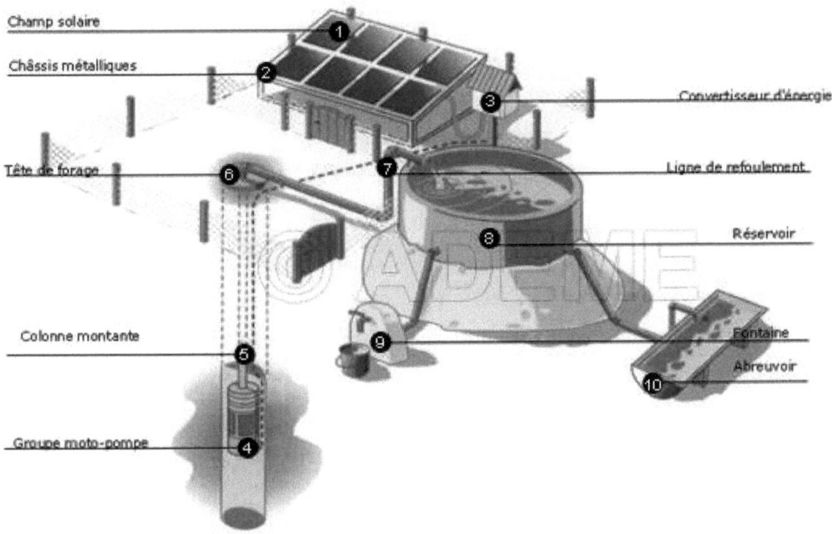

Figure 1Installation of a solar PV pumping system [5].

2.2. Photovoltaic generator

2.2.1. Photovoltaic effect

The phenomenon involved is the interaction of light with atoms. Solar radiation can be considered to be made up of particles: photons, whose energy varies with wavelength, and a junction between two semi-conducting layers (or between a metal plate and a semi-conducting layer). Each layer is connected to an electrical conductor, so there are two wires to connect the cell to an external electrical circuit.

$E_{photon} = h\nu$ where $h = 6{,}62.10^{-34} J.s$ Planck's constant and ν the frequency corresponding to the wavelength $\lambda = \frac{C}{\nu}$ and $C = 3.10^{8} m.s^{-1}$ the speed of light propagation.

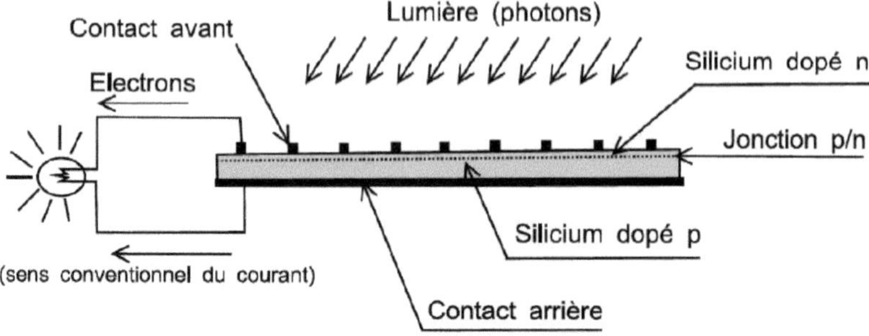

Figure 2Transformation of light energy into photovoltaic energy

2.2.2. Photovoltaic cell

The basic element in the conversion of radiation is the photovoltaic cell.

It is characterised by its low power and low voltage (0.5 V to 0.6 V).

The material most commonly used in the manufacture of photovoltaic cells is silicon, which is a semiconductor. To be usable, the electron-hole pairs must be dissociated in the volume of material. This is achieved by creating a P-N junction. Such a junction creates a zone of high electric field between zone N and zone P, at a distance from the surface chosen equal to approximately $0.3\mu m$a zone of high electric field. Any carrier, electron or hole, which has been able to diffuse to this field zone will be drained preferentially to one side (electrons to zone N and holes to zone P).

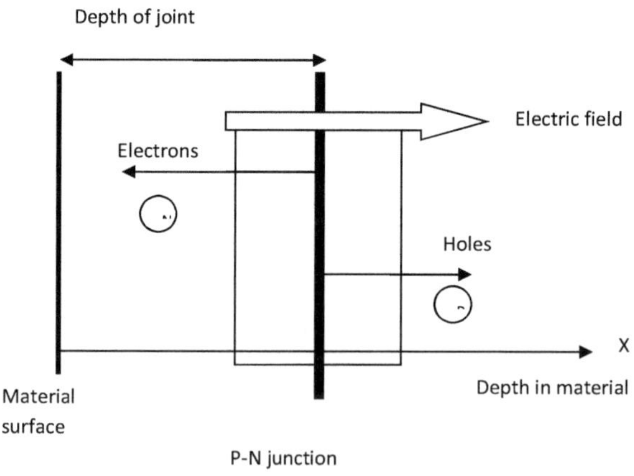

Figure 3Equivalent diagram of the P-N junction

2.2.2.1. Heliometric characteristics I = f(V)

The characteristic current-voltage curve or I = f(V) or "heliometric curve" of a solar cell or module is characterised by a relationship between the voltage U and the current I at its terminals. This relationship is shown below.

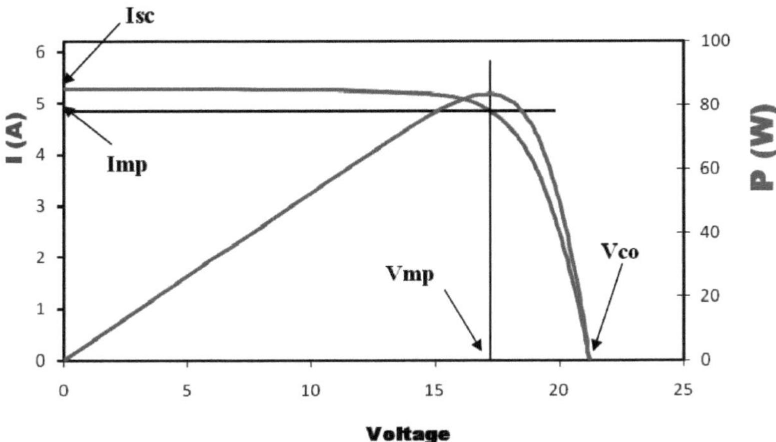

Figure 4Current-voltage characteristics of a PV cell

2.2.2.2. Modelling a photovoltaic cell

The photovoltaic cell can be modelled as follows:

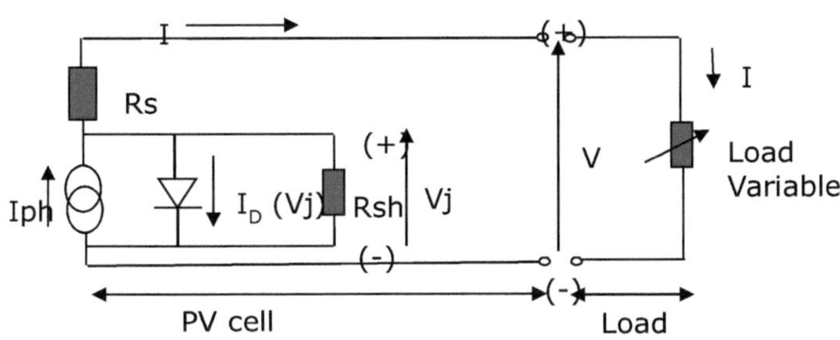

Figure 5Equivalent diagram of a photovoltaic cell

2.2.2.3. Influence of sunlight on the photovoltaic cell

For a given temperature (25°C for example) the value of the irradiation will modify the characteristic, not in its general form, but for the values I_{CC} , V_{CO} and the product (I_m p, vmp).

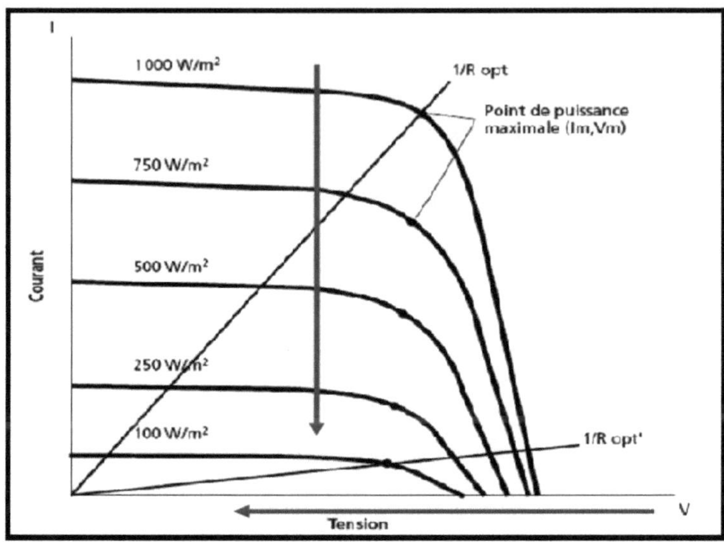

Figure 6Effect of sunlight on the PV cell

2.2.2.4. Photovoltaic cell technologies

There are several types of solar cells:

- Monocrystalline cells (15% - 22% efficiency)
- Polycrystalline cells (efficiency 10%-13%)
- Amorphous cells (5%-8% efficiency)

2.2.3. Solar module photovoltaic

Given their low electrical characteristics (power and voltage), several cells are interconnected in series (increasing voltage) and in parallel (increasing power) to form a single marketable generator called a module.

2.2.4. Solar panel photovoltaic

To obtain high power levels, modules need to be combined in series - in parallel - to form a solar panel. These are electrically connected and assembled on a frame. The size of the panel is given by the peak power. To fully define the system, we need to specify how the modules are grouped.

2.2.5. Solar field photovoltaic

A photovoltaic array is a series-parallel combination of modules designed to achieve high power levels (in excess of a hundred watts). The laws applicable to the elementary cell remain valid for the photovoltaic array.

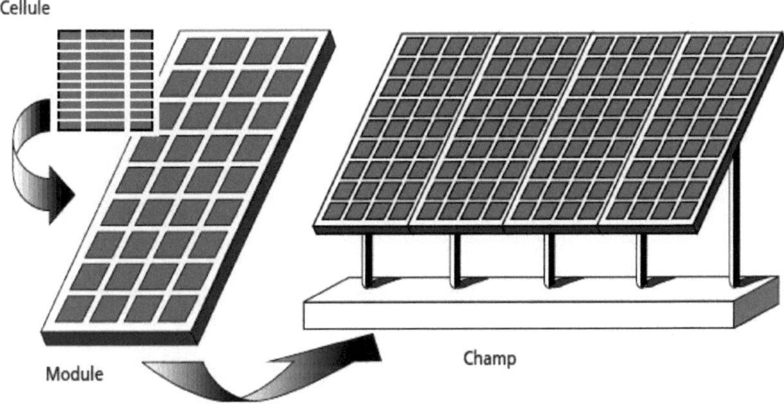

Figure 7Summary diagram of the PV array [6].

2.3. Pump unit

Pumps can be classified according to various criteria: pump design (centrifugal or positive displacement), position in the system (submerged or surface) and type of motor used (DC or AC).

2.3.1. Classification by pump design

A pump is a device used to suck in and discharge a fluid. There are two types of pump: centrifugal pumps and positive displacement pumps.

2.3.1.1. Centrifugal pumps

Centrifugal pumps use variations in the speed of the fluid being pumped to obtain an increase in pressure. The mechanical energy of the motor is transmitted to the fluid. The speed given to the fluid provides it with kinetic energy. The kinetic energy is then transformed into pressure energy.

The characteristics of centrifugal pumps are :

- The pump's drive torque is practically zero at start-up. (Particularly interesting when using photovoltaic modules, as the pump runs even when there is very little sunlight).
- There is little or no suction. They must be primed to operate so as to avoid any risk of destruction if they run dry. Some are self-priming.
- Can be submerged or surface-mounted.
- Several stages (cage + impeller) can be superimposed to obtain high pressures.

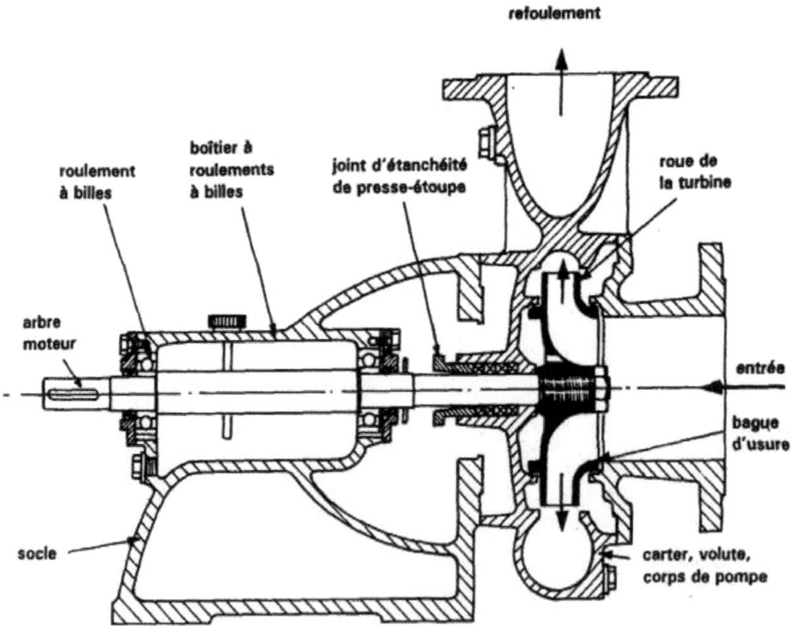

Figure 8Centrifugal pump [6].

2.3.1.2.Positive displacement pumps

Positive displacement pumps, also known as progressive cavity pumps (fig6), use variations in the volume of the fluid being pumped to obtain an increase in pressure. The fluid is first drawn in by increasing its volume and then discharged by decreasing the same volume. The most commonly used positive displacement pumps are piston, vane and gear pumps.

Their main advantages are as follows:

- They are designed for low flow rates and high heights;
- They are highly efficient, and surface pumps are self-priming;
- The starting torque of a positive displacement pump (3 to 5 times the nominal torque) ;
- The main advantage of positive displacement pumps is their ability to convey fluids at very high pressures.

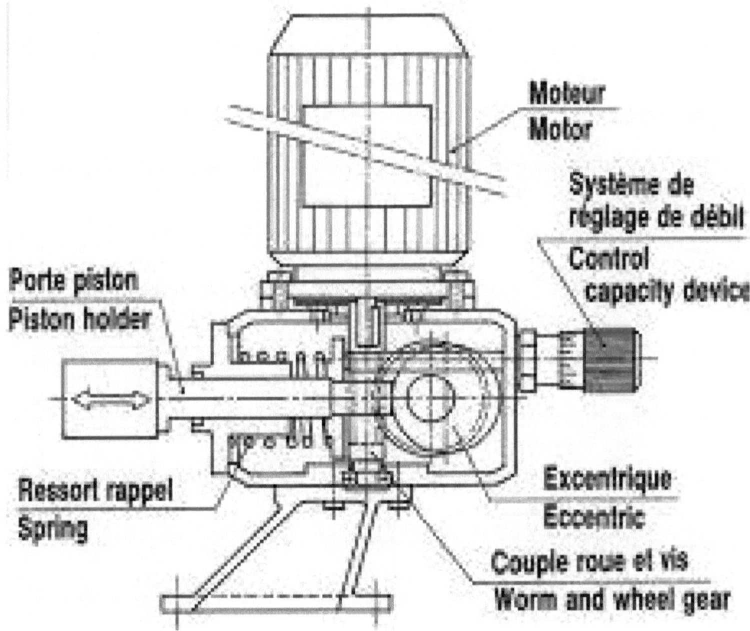

Figure 9Volumetric pump [6].

2.3.2. Classification by pump position

Depending on the physical location of the pump, we distinguish between : Surface pumps and submersible pumps.

2.3.2.1.Surface pumps

The term surface defines the position of a pump in relation to the liquid to be pumped. It is called a surface pump because it is designed to be placed outside the liquid to be sucked up.

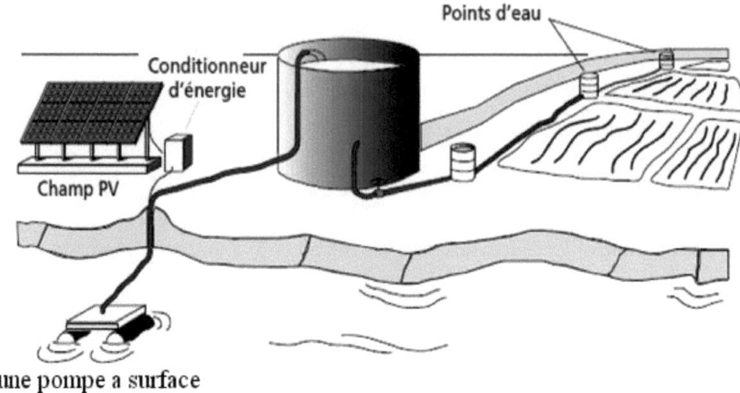

Figure 10Surface pump [6].

2.3.2.2.Submersible pumps

Discharge pumps are immersed in water and have either their motor immersed with the pump (monobloc pump), or the motor on the surface. Power is transmitted by a long shaft connecting the pump to the motor. In both cases, a discharge pipe after the pump enables the pump to be raised several tens of metres, depending on the power of the motor.

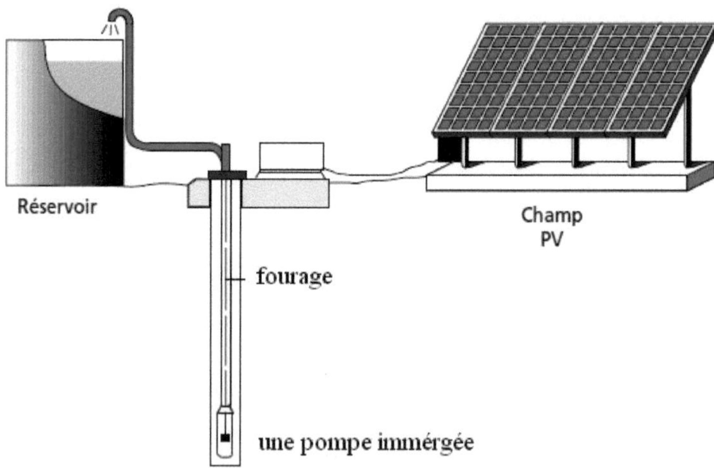

Figure 11Submersible pump [6].

2.3.3. Classification by engine used

An electric motor is an electromechanical device that converts electrical energy into mechanical energy. There are two types of motor: direct current and alternating current.

2.3.3.1. Classification by DC motor

The electrical energy applied to a motor is transformed into mechanical energy by varying the direction of the current flowing in an armature (usually the rotor) subjected to a magnetic field produced by an inductor (usually the stator). The current in the rotor of a DC motor is commutated using brushes made of carbon and graphite or by electronic commutation.

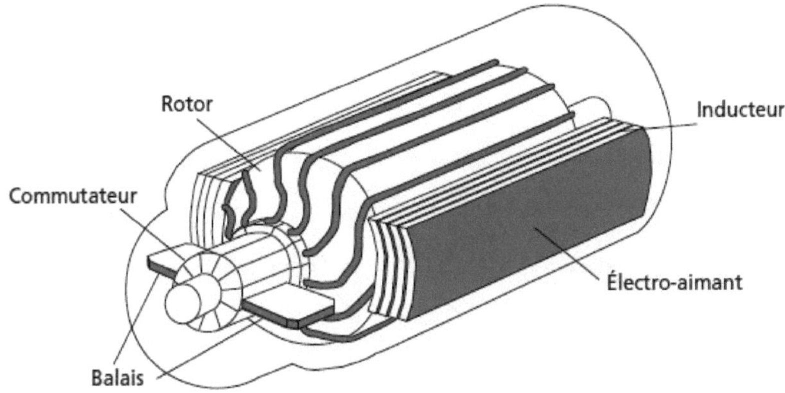

Figure 12DC brush motor [6].

2.3.3.2.Classification by AC motor

Asynchronous (squirrel-cage) AC motors are the most commonly used for a wide range of industrial applications. They are particularly used for submersible pumping in boreholes and open wells. The advent of efficient inverters has enabled this type of motor to be used in solar pumping applications.

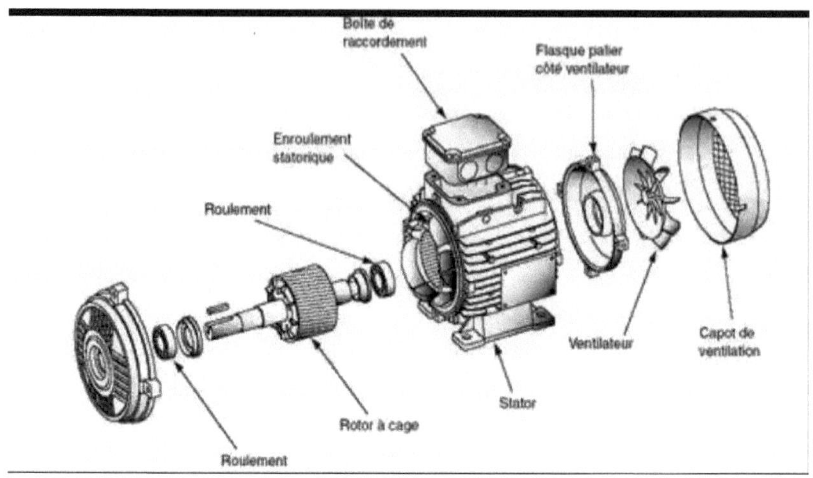

Figure 13AC motor [6]

2.4. Conversion elements

2.4.1. DC-DC converter (chopper)

Used in the case of a pump mounted on a DC motor. Otherwise the torque is direct.

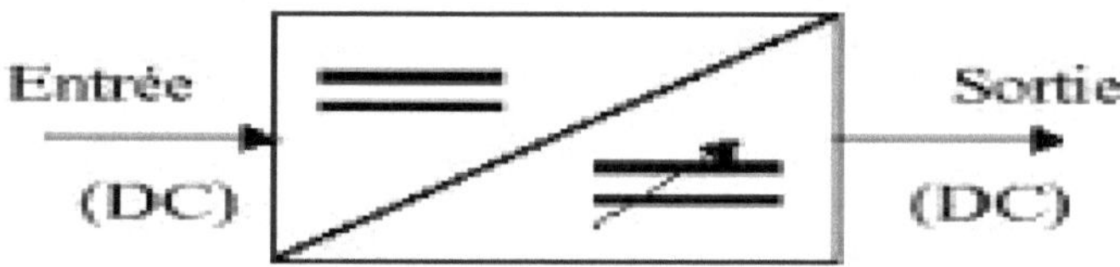

Figure 14DC-DC converter [6].

2.4.2. DC-AC converter (inverter)

The main function of the inverter is to transform the direct current produced by the solar panels into alternating current. The variable-frequency inverter transforms the direct current from the panels into single-phase or three-phase alternating current, which is used to operate the pump. This transformation is carried out with excellent efficiency, in excess of 95%.

The frequency of the output current (pump speed) is variable so as to adapt the power absorbed to that supplied by the panels. In addition, a microprocessor-controlled device ensures that the panels are always used at maximum power.
Lastly, the inverter is fitted with a range of protections (overheating, overcurrent, low water level in the borehole, etc.) and a shutdown system when the reservoir is full. Its robust, watertight construction and choice of tried-and-tested components make it highly reliable under the toughest operating conditions.

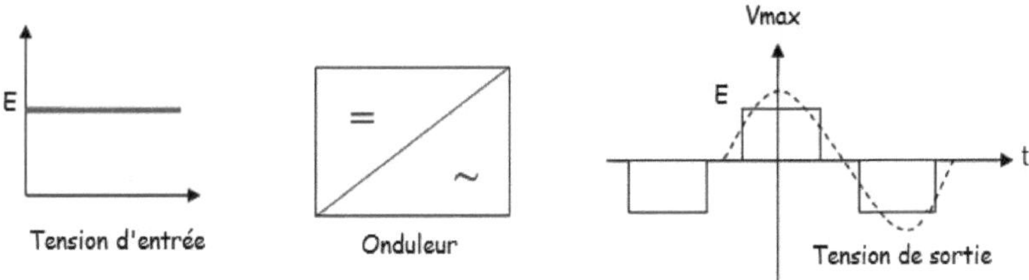

Figure 15DC-AC converter [6].

2.5. Storage area

In solar pumping systems, energy can be stored in one of two ways: electrical energy or water. The latter method is often adopted because it is more practical to store water in reservoirs than electrical energy in heavy, expensive and fragile accumulators, and energy efficiency is better when there are no accumulators.
The energy supplied by photovoltaic generators is expensive, despite the falling cost per watt-peak. It is therefore necessary to operate these generators at their optimum power.

Conclusion

In this chapter we have presented the essential components of a solar PV pumping system and their main functions, each in its own right, which has led us to understand that: Solar photovoltaic pumping remains the most widely used technology because it uses reservoirs for storage instead of batteries and this can reduce investment costs. Given the roles that these elements play in this system, it is essential to dimension them whenever possible before designing such a system, which will be the subject of the next chapter of this dissertation.

CHAPTER 3: SIZING A SOLAR PUMPING SYSTEM PHOTOVOLTAIC

Introduction

Sizing a photovoltaic system means determining all the elements in the photovoltaic chain to guarantee a good supply of energy, based on stresses such as sunshine and the load profile. In all cases, you need to know the water requirements and solar energy potential of the location concerned. This enables you to make the right choice of photovoltaic modules, their layout and support structure, and the choice of electrical components to regulate and protect the system and its users.

In this chapter we will look at the methodology for sizing a solar PV pumping system. The general procedure will make it possible to roughly size the components of a pump in order to give an order of magnitude of these components.

3.1. Methodology

A pumping system can be designed either analytically or using software.

3.1.1. Analytical methods

Analytical methods are those for which all the calculations and selection of elements can be carried out manually, taking into account the main parameters as well as the information available on the study site.

3.1.2. Methods using software

The software method is not necessarily complex. It is a method for optimising calculations. It can be applied whatever the size of the system.

3.2. State of the place

The study site concerned is located on the CNES premises. It is a moringa field spread over a rectangular plot 49 m (forty-nine metres) long and 22 m (twenty-two metres) wide, giving a surface area of 1078 m^2 (one thousand seventy-eight square metres). The premises already have a well drilled to a total depth of 45 m (forty-five metres), a plastic water storage tank placed at a height of 5 m (five metres) above the ground on a concrete slab support 3 m (three metres) high, and dowels. The distance between the well and the tank is 25 m (twenty-five metres).

The job involves designing a solar PV pumping system on the site so that the field can be irrigated using drip irrigation.

Photo 1Field to be irrigated

Photo 2Storage tank

3.3. Drip irrigation

Drip irrigation is also known as localised irrigation or micro-irrigation.

". It is called "infiltration" irrigation when it is carried out using buried filter pipes. Drip irrigation is becoming increasingly popular as a means of coping with water scarcity. It is characterised by a localised, frequent and continuous supply of water using low flow rates at low pressures. Only the part of the soil used by the roots is continually moistened. This reduces evaporation, preserves soil structure and reduces weeds. This system can also be used on fields with irregular topography and configuration, heavy soils that crack in summer, or light, filtering soils. The high frequency of watering dilutes the salts present in the soil solution under the distributor and keeps them at the periphery of the wetted bulb.

3.4. Dimensioning using the analytical method

There are two main objectives in sizing the solar system we are studying here: saving water and saving money. In other words, a properly sized photovoltaic system for pumping water for drip irrigation must be able to meet the energy requirements expected of it, while being the least costly both at the time of installation and during the production phase, and must operate for a reasonable period to guarantee its amortisation.

For this reason, the following steps should be followed to achieve optimum sizing:

- ✓ Determining the water requirements of the field to be irrigated ;
- ✓ Calculating the hydraulic energy required ;
- ✓ Sizing and selecting the type of motor-driven pump ;
- ✓ Estimated available solar energy ;
- ✓ Sizing the PV generator.

3.4.1. Determining the water requirements of the field to be irrigated

For drip irrigation, water is poured onto the plant as frequently as possible or continuously, and on a daily basis. As for determining the quantity of water to be supplied (Q), irrigation must be carried out in such a way as to compensate for losses through evaporation and prevent salinisation of the rhizosphere. Moringa, which is classified as a market garden crop, has an estimated water requirement of 60m^3 /day/ha, as shown in the table below.

Table 1Water requirements

Irrigation	Flow rate Q(m^3/day/ha)
Vegetable farming	60

Rice	100
Seeds	45
Sugar cane	65
Cotton	55

For this study, our field covers an area of 1078 m², i.e. 0.1078 ha.

As a result, the water requirement for our field is :

$$Q = 60 \times 0{,}1078 \qquad (2.1)$$

Q= 6.468 m^3 /day; Q≈ 6.5 m^3/day

3.4.2. Calculation of hydraulic energy required

Once the volume of water required for each month of the year and the characteristics of the well have been defined, we can calculate the average daily and monthly hydraulic energy required using the relationship :

$$E_h = C_h \, x \, Q \, x \, H_{MT} \qquad (2.2)$$

With :

E_h : Hydraulic energy (Wh/d)

C_h: Hydraulic constant (kg.s.h/m²), dependent on the earth's gravity and water density

C_h = g×ƒ/3600 = 9,81× 1000/3600 = 2.725 (kg.s.h/m^2)

Q: Daily flow rate (m^3)

HMT: Total manometric height (m)

3.4.2.1. Total head

The total head of a pump is the pressure difference in metres of water column between the suction and discharge ports. It is determined by the following relationship

$$H_{MT} = H_g + \Delta J \qquad (2.3)$$

Hg: Geometric height (m) ;

$$H_g = N_d + H_r \qquad (2.4)$$

$$H_{MT} = N_d + H_r + \Delta J \qquad (2.5)$$

With :

N_d : Dynamic level (m)

H_r : Tank height (m)

Δj Pressure loss: represents the pressure losses produced by the friction of water on the pipe walls. These losses correspond to 10% Hg.

In this case :

N_s = 11 m

N_d = 22 m

H_r = 05 m

Δj = 2.7 m

$H_{MT} = 22 + 5 + 2.7 = 29,7\ m$ $H_{MT} \approx 30$ m

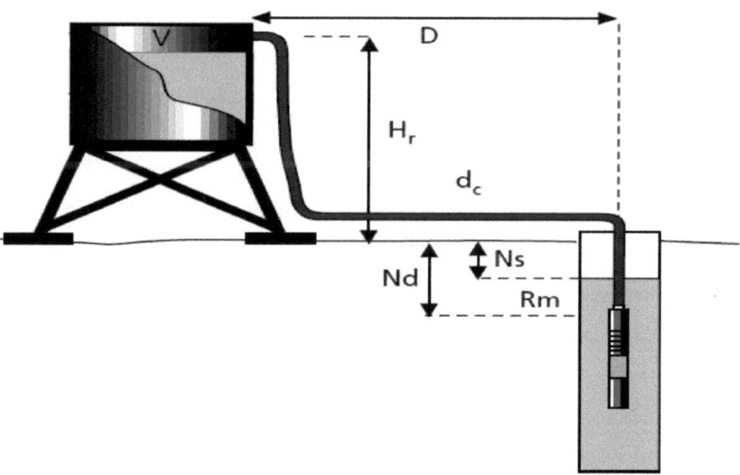

Figure 16Total Manometric Head (TMH) [8].

Finally, we have the hydraulic energy corresponding to :

$E_h = 2{,}725\ \times 6{,}5\ \times 30 =\ 531{,}375\ Wh/j$

E_h = 531.375 Wh/d

3.4.3. Sizing and choosing the type of motor-driven pump

The choice of pump depends on the hydraulic characteristics of the proposed installation (flow rate, head). Figure (3.2) shows a classification of pumps according to total head and required flow rate.

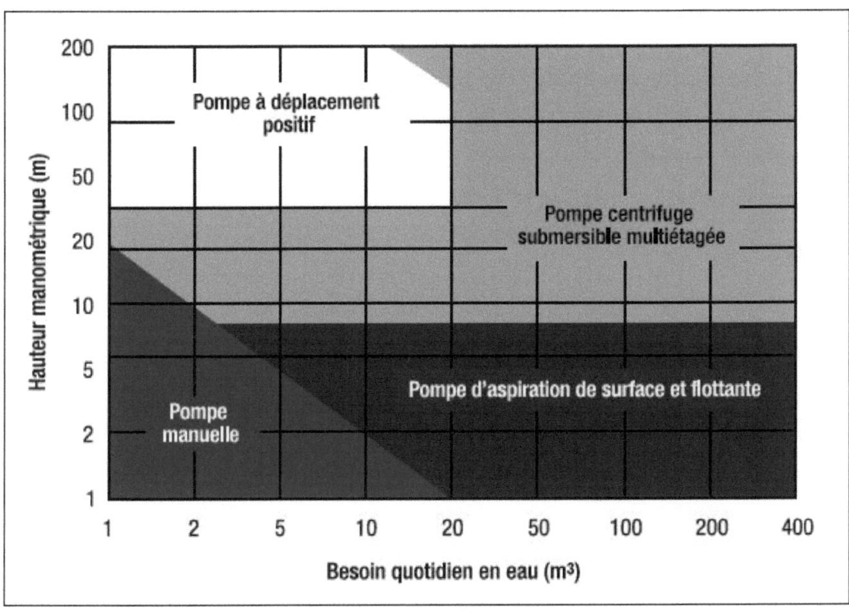

Figure 17Pump selection based on HMT and flow rate [5].

For our flow rate (6.5 m^3/d) and the HMT (30 m), this figure shows the choice of a positive displacement pump. But for the ideal choice of pump for our system, it is also chosen according to the HMT and the hourly flow rate, which is the quotient of the daily flow rate over the maximum duration of sunshine (8h for this study

$$Q_h = \frac{Q}{T} \qquad (2.6)$$

$Q_h = \frac{6.5}{8} = 0{,}8125\ m^3$/h, i.e. 812.5 l/h

Hence our choice of the Lorentz pump model PS200-HR-07 in the table below.

Table 2Lorentz PS200 pump range

PS200	HR-04	HR-07	HR-14
HMT(m)	0-50	0-30	0-20
Max flow (m^3/h)	0.8	1.2	2.7
Direct solar operation	24-48VDC	24-48VDC	24-48VDC
Yield (%)	40-60	40-61	40-62

3.4.4. Estimating available solar energy

It is also known as electrical energy. The energy required to lift a given quantity of water to a given height over the course of a day is calculated using the following equation:

$$E_{elec} = \frac{E_h}{R_{mp} \times R_{ond}} \qquad (2.7)$$

With :

E_{elec} : Electrical energy (Wh/d)

E_h : Hydraulic energy (Wh/d)

R_{mp} : Pump unit efficiency

R_{ond} : Inverter efficiency

When the torque is direct (without inverter) the previous expression becomes the following:

$$E_{elec} = \frac{E_h}{R_{mp}} \qquad (2.8)$$

$$E_{elec} = \frac{531.375}{0.4} = 1\ 328{,}5 \text{ Wh/d}$$

E_{elec} = 1 328.5 Wh/d

3.4.5. Generator sizing photovoltaic

This sizing takes into account (roughly) the rated power of the module array (P_c) plus the various losses (voltage conversion, regulation, etc.) and the amount of sunshine on the site. Its installation (orientation) depends on its geographical position (latitude). The actual sizing, i.e. the series-parallel arrangement of the solar panels, is determined in conjunction with the choice of pump to ensure compatibility between energy supply and demand. We have seen that the voltage and current characteristics of the photovoltaic modules and the pump must be correlated. If the correlation is not respected, some elements may break down. All of this is scrupulously regulated by the inverter if necessary (in the case of a pump operating on alternating current).

The site is located at latitude 13.52° North, a longitude of 2.10° East and an altitude of 207 m, corresponding to the location of the town of Niamey (Niger), giving an optimum orientation of 15° North-South for the attachment of the solar modules.
As far as sunshine is concerned, we have opted for the annual average, while taking into account the sunshine in the worst month of the year.
This annual average is around 5.41 kWh/m²/day (see appendix), especially for our site.
To determine this peak power, we used relation (2.7) and relation (2.9) below:

$$E_{elec} = P_c \, x \, H_i \, x \, K_p \qquad (2.9)$$

With :

P_c : peak power (W)

H_i : Solar irradiation (kWh/m²/d)

K_p : generator productivity coefficient, $K_p = 0.8$ for a system without batteries

Equating relations (2.8) and (2.9) we get :

$$P_c = \frac{C_h \, x \, Q \, x \, H_{MT}}{R_{mp} \, x \, H_i \, x \, K_p} \qquad (2.10)$$

From which we deduce that :

$$P_c = \frac{E_{elec}}{H_i \, x \, K_p} \qquad (2.11)$$

$$P_c = \frac{1\,328{,}5}{5{,}41 \times 0{,}8} = 306{,}94 \text{ WC}$$

$$P_c \approx 307 \, W_c$$

3.4.5.1. Number of panels

The number (N) of panels required for this installation is determined as follows:

- $N_{sr} = \frac{V_{syst}}{V_{mod}}$ (2.12)

With :

N_{sr} : Number of panels in series ;

V_{syst} : System voltage ;

V_{mod} : Solar module voltage.

$$N_{sr} = \frac{36}{12} = 3 \, ;$$

- $N_p = \frac{P_c}{N_s \times Pc_{mod}}$ (2.13)

With :

N_p : Number of panels in parallel ;

Pc_{mod} : peak power of the solar module, shown in table (2.3) below

$$N_p = \frac{307}{3\times110} = 0,93 \qquad N_p \approx 1$$

Table 3Characteristics of the solar module used

Technology	Sun Plus
Peak power	110 Wp
Vmp	18 V
Imp	6.5 A
Voc	21.6 V
Icc	7 A

Hence the installed peak power (P_{ci}) is given by :

$$P_{ci} = N_{sr} \ x \ N_p \ x \ P_{cmod} \qquad (2.14)$$

$$P_{ci} = 3 \times 1 \times 110 = 330 \text{ WC}$$

3.4.6. Sizing cable and pipe cross-sections

- The cable cross-section is determined by the following formula:

$$S \geq \frac{2 \ x \ \varphi \ x \ L \ x \ I}{\Delta V \ x \ U}$$

With ; φ resistivity, which is 0.017 Ω/mm²/m for aluminium conductors ;

L: length 25 m; I: generator current (A); U: generator voltage (V) ;

ΔV Voltage drop, it of 3% according to French standard C 15712

$$S \geq \frac{2\times0,017\times25\times5,56}{0,03\times54} \geq 2,91 \ mm^2$$

We have chosen a standard cable cross-section of 6 mm², taking into account an overcurrent in the event of a short-circuit.

- The diameter of the piping was chosen on the basis of the technical data sheet for the pump used, which recommends a diameter of 4 inches (4"), i.e. 100 mm.

In summary, this method gave us the following results:

Table 4Results of the analytical method

Water requirements	6.5 m^3/d
HMT	30 m
Hydropower	531 .375 Wh/d
Electrical energy (solar)	1,328.5 Wh/d
Peak power	307 Wc
Number of panels	3
Installed peak power	330 Wp
Generator voltage	54 V
Generator current	6,5 A
Cable cross-section	6 mm2

3.5. Dimensioning using software

3.5.1. PVsyst software Version 6.7.3

PVsyst 6.7.3 is a software package for simulating and sizing stand-alone, grid-connected and pumped solar photovoltaic installations. The software was developed by the University of Geneva (Switzerland), and was designed by André Mermoud.

The PVsyst 6.7.3 software has several inputs: average monthly solar flows, average monthly temperatures, water requirements, choice of PV modules and their inclination, choice of pump and its control system, input of the number of days of autonomy, solar coverage rate and investment cost (purchase of equipment, system installation cost).

Photo 3Screenshot of the PVsyst home page

3.5.2. Dimensioning stages

- Enter the basic data (country, latitude, longitude, altitude, time zone) ;
- Orientation (inclination of collectors, annual, seasonal or monthly irradiation) ;
- User requirements (borehole and reservoir characteristics) ;
- System (autonomy, choice of modules, pump and control system) ;
- Simulation

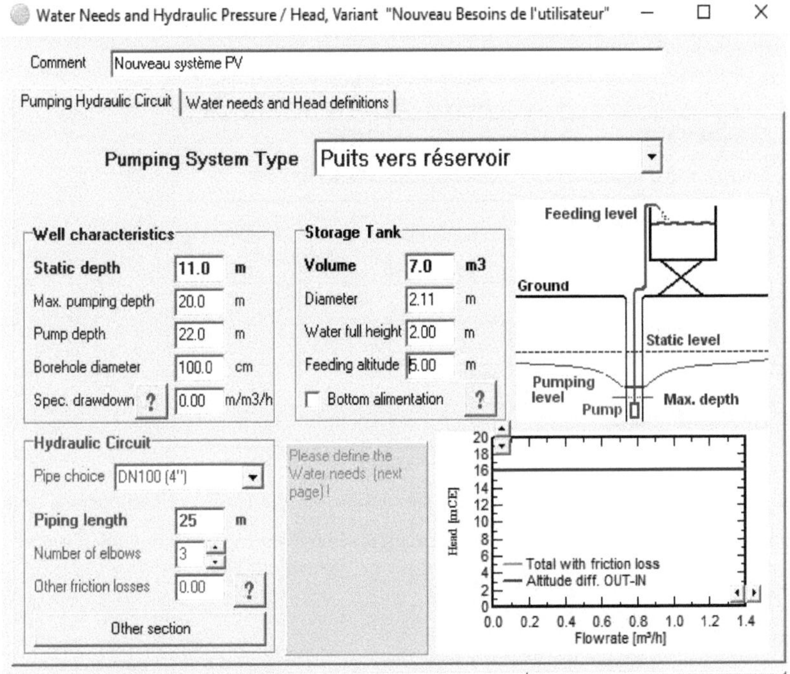

Photo 4Screenshot of the PVsyst solar pumping system

Simulation results

The software used to obtain our results is an evaluation version (30 days).

The main results of the simulation are: water pumped, water requirement, water shortage, energy supplied to the pump, unused energy, specific energy, system efficiency, peak power of the field, number of **modules**, surface area of the field of modules shown in the table below:

Table 5PVsyst simulation results

Pumped water	2 373m^3/year
Water requirements	2 373m^3/year
Missing water	0,0%
Energy at the pump	341 kWh/year
Unused PV energy (full tank)	141 kWh/year
Specific Energy	0.13 kWh/m^3
System efficiency	64,6 %
Overall field power Nominal	255 Wp
Total number of modules	3
Total module area	1,9 m^2

This table summarises the results of the simulation, for the choice of components (pump, module, control device) and their characteristics, see the appendices (simulation report).

Conclusion

At the end of this chapter, we have developed the sizing steps of a solar PV pumping system in two methods, an analytical method and a software method, which are applied to a real case. All these methods require different criteria for the choice of components.

The results obtained in this section are specific to the CNES solar PV pumping system designed to supply a 0.1078 ha field with a water requirement of 6.5 m^3/day.

However, installing and setting up such a system requires a certain investment cost for ideal operation, which will lead us to carry out an economic study of the system in the next chapter.

CHAPTER 4: ECONOMIC STUDY

Introduction

The economic analysis of photovoltaic systems (pumping, lighting) is becoming increasingly important now that this technology is sufficiently mature to compete with conventional systems (interconnected grid, diesel). This economic analysis is essential for making informed investment decisions, comparing forecasts with the reality of projects and programmes, quantifying the profitability of the services provided by the system and motivating decision-makers, investors and potential users.

4.1. Provisional budget for the analytical method

To do this, we have given the price of the various components in the table below in order to estimate a total investment cost.

Table 6Investment costs 1

N°	Components	Unit price (F CFA)	number	Total price (F CFA)
1	Solar module	90 000	3	270 000
2	Fixing the modules	12 500	3	37 500
3	Pump	1 200 000	1	1 250 000
4	Piping (55 m)	6 000	55	330 000
5	Cable a roller (4× 6 mm^2)	5 000	100	500 000
6	Civil engineering and accessories	900 000	1	900 000
7	Total investment cost			3 287 500

4.2. Provisional budget for the software method

This part has been compiled in the software itself, which has an economic evaluation stage just before the simulation, which gives the following result:

Table 7Investment costs 2

N°	Components	Unit price (F CFA)	number	Total price (F CFA)
1	PV modules (Pnom = 85 Wp)	75 000	3	225 000
2	Support and integration	12 500	3	37 500
3	Pump (Pnom = 190 W)	1 000 000	1	1 000 000
4	Regulator, Converter	55 000	1	55 000
5	Construction, wiring	5 000	100	500 000
6	Piping	6 000	55	330 000
7	Engineering	900 000	1	900 000
8	Total investment cost			3 047 500

4.3. System reliability

4.3.1. System maintenance

Maintenance is the set of actions that enable an asset to be maintained or restored to a clearly specified state; to maintain well is to ensure that these operations are carried out at minimum overall cost. The main maintenance activities are :

- prevention: visits, inspections ;
- introduction: revisions, repairs ;
- improvement: renovations, modernisations.

Maintenance, whether preventive (systematic or conditional) or corrective (troubleshooting or repair), must be closely linked to system operation.

However, this system does not necessarily require a great deal of maintenance. The main maintenance is carried out on the solar modules and the pump, as follows

- ✓ The modules need to be cleaned regularly (in the evening or early morning every day or 3 or 4 times a week) because of the dust.
- ✓ Check that the solar panel is not partially shaded.
- ✓ If a module fails, it is replaced by another identical module.
- ✓ Regular cleaning of the pump is necessary at least every two years because of the sludge.

4.3.2. System lifetime

The following table shows the service life of the various components of our photovoltaic pumping system:

Table 8Component service life

Components	Service life
Solar module	25 years old
Fixing the modules	25 years old
Pump	18 years old

4.4. Economic profitability study

It is interesting to evaluate such systems to see how profitable they are over the long term. This is a practical reflection of their importance. To do this, we evaluate the total cost of using a diesel motor-driven pump and its amortisation in order to estimate the potential savings by replacing the conventional pumping system with a solar photovoltaic pumping system.

We chose the Yamaha YP 20C motor-driven pump, which has a service life of 2 years at 2 hours a day and a maximum flow rate of 36 m^3/h.

Table 9Cost of using the motor-driven pump for 2 years

Designation	Price F CFA
Motor pump	320 000
Fuel consumption (2l/d)	788 400
Drain every 2 weeks	78 000
Maintenance and repairs	50 000
Total cost	1 236 400

If we look closely, the cost of using the classic motor-driven pump for 2 years is one million two hundred and thirty-six thousand four hundred CFA francs (1,236,400 F CFA). So using the motor-driven pump for 6 years will cost three million seven hundred and nine thousand two hundred CFA francs (3,709,200 CFA francs), which is even more than the investment in our photovoltaic system (3,287,500 CFA francs), which has a minimum lifespan of eighteen years (18 years).

When we calculate the cost of the motor-driven pump for a period of 18 years, it will be eleven million one hundred and twenty-seven thousand six hundred CFA francs (11,127,600

F CFA), whereas the solar photovoltaic pump will cost three million two hundred and eighty-seven thousand five hundred CFA francs (3,287,500 F CFA) **for the** same period. Here, by using solar pumping, the farmer will benefit from a free pumping system for 12 years, unlike the one who uses the motor pump, which will save him a sum of seven million eight hundred and forty thousand one hundred CFA francs (7,840,100 F CFA) compared to the one who used the motor pump for 18 years.

Conclusion

The efficiency of any system depends on its cost and lifespan. The purpose of the study presented here is therefore to determine the price of all the components of our system in order to estimate the total cost of the system (capable of supplying the CNES field), which amounts to a maximum of three million two hundred and eighty-seven thousand five hundred CFA francs (3,287,500 CFA francs).

Admittedly, the cost is a little high for a start-up, but when you take into account the low maintenance required by the system and its lifespan before amortisation (18 years minimum) compared with the use of a conventional motor-driven pump, they make our system: efficient, reliable and economically profitable.

General conclusion - Recommendations

Global demand for energy is changing rapidly, and natural energy resources such as uranium, gas and oil are diminishing as a result of the widespread use and development of industry in recent years. To cover energy needs, research is being carried out into renewable energy. One of the renewable energies that can meet demand is photovoltaic solar energy, which is clean, silent, available and "free". This explains why its use is growing significantly around the world.

This thesis focused on the study of a photovoltaic solar pumping system. The system studied can be considered an ideal solution for supplying drinking water to sparsely populated and isolated regions, because it is simple to install and can be adapted to meet a variety of energy needs, and the operating costs are acceptable given the low maintenance requirements. What's more, photovoltaic energy is totally scalable, so it can meet a wide range of needs. The size of the installation can also be subsequently increased to meet the needs of its owner. In the case study, the work consisted of sizing a solar PV pumping system to irrigate a field of 0.1078 ha **of** moringa using drip irrigation located on the CNES premises in the town of Niamey. This led us to present and size the main components of the system one by one. Whatever the sizing method, it is necessary to have accurate historical and current meteorological information, both for calculating the water requirements of the field to be irrigated and for calculating and sizing the photovoltaic array.

However, it is important to point out that calculating the size of the generator often involves a degree of uncertainty. There are two main reasons for this uncertainty: the first relates to the random nature of solar radiation, which is often not well known. The second is linked to the difficulty of estimating the water requirements. It is therefore advisable to take precautions when choosing the type of pump and the size of the generator.

Finally, an economic study was presented to determine not only the cost of the system to be installed at the CNES but also to assess its profitability compared with the use of the diesel motor pump. It is therefore important to remember that such a system can increase the rate of access to drinking water, particularly in rural areas, and substantially increase national agricultural production through the development of small irrigated agricultural areas.

Recommendations

Our recommendations are as follows:

- **For the host organisation**
  - Set up a technical team to supervise the PV system and the drip line;

- Increase the volume of storage capacity ;
- Set up a wifi network within the structure to facilitate document searches, especially for the Research and Engineering departments.

❖ **For farmers**

- Replace the conventional pumping system with solar PV pumping

❖ **For the State of Niger**

- Promoting such a project at national level could lead to a significant increase in national agricultural production with a view to achieving food self-sufficiency;
- Subsidising the price of all solar equipment to make it available and accessible to everyone, in order to bridge the energy gap and take advantage of this technology, which will undoubtedly be the energy of the future.

Bibliography

[1]. CNES archives 2017-2018

[2]. M. Samey Manou, Feasibility study for the construction of an autonomous solar PEA water supply station in Chinaoua Haoussa, rural commune of Ourno, Madaoua department, Tahoua region; 2002

[3]. Thesis: Ilyass Boudouar and Khalid Melouard (May 2017); Sizing and installation of a photovoltaic pumping system, Ibn Zohr University - Agadir (Algeria)

[4]. Dissertation: lary Ligring (October 2012); Study for the implementation of the

Solar pumping of a nine-hectare field for drip irrigation in seheba, au

Chad, International Institute for Water and Environmental Engineering (Burkina Faso)

[5]. Dissertation: Thierry Maurice (February 2007); Photovoltaic system: sizing for water pumping, for drip irrigation, University of Ouagadougou (Burkina Faso)

[6]. Thesis: Degla Mohammed Larbi and Ben Ahmed Bachir (May 2017); Dimensioning of a photovoltaic pumping system, Université Kasdi Merbah Ouargla (Algeria)

[7]. Dissertation: Amal Resfa (June 2007); Study of a photovoltaic pumping system, Aboubakr Belkaïd University (Algeria)

Webography :

[8]. www.africarriatenergie.com; consulted in May 2018

[9]. www.lorentz.de; consulted in June 2018

[10]. www.pvsyst.com; consulted in July 2018

[11]. https://www.memoireonline.com; consulted in July 2018

Table of contents

Printed by Books on Demand GmbH, Norderstedt / Germany